T0156439

Bird from Hell

And Other Mega Fauna

Gerald McIsaac

Order this book online at www.trafford.com
or email orders@trafford.com

Most Trafford titles are also available at major online book retailers.

Printed in Victoria, BC, Canada.

ISBN: 978-1-4269-2305-0 (Soft)
ISBN: 978-1-4269-2306-7 (Hard)
ISBN: 978-1-4269-2580-1 (e-book)

Library of Congress Control Number: 2009914169

Trafford rev. 12/31/2009

www.trafford.com

North America & international
toll-free: 1 888 232 4444 (USA & Canada)
phone: 250 383 6864 • fax: 812 355 4082 • email: info@trafford.com

Chapter I

Devil Bird

Here in the mountains of western North America, after the sun goes down, Satan opens the gates of hell and turns loose his pet, the Bird from Hell. In the darkness, the devil bird hunts. My First Nation friends assure me that this animal is quite capable of killing big game such as moose and horses, but its preferred prey is smaller animals such as dogs and deer. Then too, it also preys on humans. You see, this animal is a man eater.

Since I am not the terribly shy sort, I asked the girls in the village to describe this elusive monster. They said that it had the head of an eagle, a body like an anaconda, a wing span as wide as a house and a tail as long as a man is tall. The tail also ends in the shape of an arrowhead.

It is fortunate that I was sitting down when they gave me that description. It saved me from falling down. I realized that they were giving me a detailed description of a pterodactyl.

That is not the only animal they described in great detail. The other species they described include the woolly mammoth, the plesiosaur, the cave bear, which is also known as the short faced bear or the mega bear, the dire wolf and the largest ape ever to walk the earth, Gigantopithecus. That ape is commonly referred to as Bigfoot or Sasquatch, and not only is it not extinct, it is a separate species of human, living among us here in North America.

I have been in touch with government agencies and notified them of the existence of these species. As they put it, they are sceptical. They were being polite, of course. In fact, they do not believe a word I am saying. At least, they are hoping and praying I am mistaken. It would never occur to them to offer any assistance whatsoever in proving the existence of these animals. The scientific experts have too much to lose. Their reputations are at stake. The experts in this particular field, whatever that field is called, are not at all anxious to be proven the incompetent fools that they are. If that sounds unduly harsh, consider the facts. There have been countless eyewitness sightings of these animals, all of which have been ignored by the authorities. Their minds are made up and they do not want to be confused by the facts. The trouble is that facts have a rather annoying way of getting in the way of our beliefs, and the facts are that these species are right here in the mountains of North America. The scientific souls have never seen them for the very fine reason that they have never looked. What is more, they have no intention of looking. Once these various species are proven to exist, then countless careers will be ruined.

For the last several years I have been working on this science project, as I call it. Whenever possible, I have received first hand accounts from eye witnesses. Of course, this has not always been possible. Some people have moved, and others have died. The eyewitness accounts agree with the reports that the Elders have given me. I might add that the Elders are the absolute authorities on these mountains. I stress that no one whatsoever knows more about these mountains than the First Nation Elders.

The response of the scientists to my research is not too surprising. They say that I should prove these assertions. Clearly, they expect me to do the job which they are not capable of performing. This merely proves their incompetence. So be it. As yet, I do not have any smoking gun, so to speak, even though I have spent countless hours on this project. Most of that time has been spent gathering information. As they say, know your enemy. Knowledge is power. It is also true that I have to work for a living, so my time and certainly my money is limited. But since the scientists insist that I must prove that which they are not capable of proving, I am determined to do just that. That is part of the reason that I am going public with this information. At the moment I am focused on proving the existence of the pterodactyl, and I most certainly will. It is just a matter of time. The trouble is that time is not on our side, since this animal is a man eater and the public is not aware of the existence of this reptile. As a result, people are disappearing on a regular basis from open areas, such as highways. This has got to stop.

I am sure that there are people who are interested in helping to prove the existence of these species, and feel free to join the club. Do not expect any help from any government agency or scientific group. They will regard you as a threat to their security, just as they regard me as a threat. And in fact as these various species are proven to exist, they will be scrambling to find excuses for their incompetence. They are not about to admit that they have no excuse, and they certainly cannot say that they were not warned. I am sure the excuses they will soon be coming out with will be quite creative.

Perhaps a little background check is in order, just to prove my qualifications, so to speak. I have lived with First Nations people since the nineteen seventies, here in the mountains of north central British Columbia. When I first arrived we were settled on the banks of the Ingenika River, and the luxuries were few and far apart. Such items as electricity and running water were out of the question. Fresh meat was available, but it first had to be shot. That was the fun part, while the hard part was packing the meat to the cabin. We all lived in log cabins in those days, and while life was hard, we certainly enjoyed ourselves.

We now have a new village located at the north end of Williston Lake, which is a huge man made lake. The nearest town of Mackenzie, which is a metropolis of about four thousand people, is located at the south end of the lake.

Now things are different and we live in a modern village complete with electricity and running water. The village has a set of large generators which provide electricity for the all the homes, and there is a large water tower which provides us with drinking water. All the new houses are modern and we have a fine school and store. The trouble is that some of us still miss the log cabins that we used to live in. There is a price to be paid for progress.

The new and modern village was built in the nineteen nineties, and more than that, the roads were put in place connecting this village with civilization. In the seventies there were no roads up here, and the only way in or out was by aircraft or boat, Of course, one could always walk, but since the nearest town was three hundred fifty kilometres or two hundred miles on roads which did not exist, that was a tall order. The mail was flown in twice a month, and in the summer the barge brought in the supplies. It was always best to stock up in the fall, or just face the reality of going without.

For those of you who are not familiar with the mountains, be advised that we live in something called the Rocky Mountain Trench. That means we live between two mountain ranges, the very young and therefore very rugged Rocky Mountains to the east and the older and therefore more worn down mountains to the west. This range of mountains runs from the Arctic to Mexico, if I am not mistaken. It makes for beautiful scenery and a rather rough life. These mountains tend not to be very forgiving. A couple girls nearly lost their lives last winter when they broke down on the road to town. They spent a very cold night in the vehicle, and were fortunately found the next day. They would not have survived another night in the forty below weather.

Of course the reason they were not found was that no one was looking for them, and there is very little traffic on that road. The first vehicle that came along stopped and assisted those girls. What I mean is that the roads in this place are not to be confused with highways.

That is life in the mountains, and while we now have more luxuries, it is still a dangerous place to live. It is still no place for a pilgrim and any moment of complacency can prove fatal. In the seventies the mail was flown in twice a month, while now we have mail flown in three times a week. How times do change. We now even have a post office in the village, which was just set up last summer. Even though the place is becoming downright civilized, it has yet to go completely to the dogs.

There are currently about one hundred and fifty people in the village, mainly First Nation but a handful of Caucasian mountain men such as myself. The people here are Dene, and the village is called Tsay Keh Dene. These days we generally refer to it as T. K. Loosely translated, that means Mountain People, Let there be no mistake, they are mountain people. They are the experts on these mountains and the animals that live in them. I have no doubt whatsoever that the stories they are telling me are true, and that the descriptions they are giving me of these animals is detailed and accurate.

The implications are staggering. I have spent the last several years making every effort to verify the accuracy of these descriptions. Assuming my First Nation friends are correct, and I am convinced they are, then the scientific community has made some very serious errors.

According to the scientific experts this animal, the pterodactyl has been extinct for sixty five million years. That at least is the opinion of many scientific experts, who maintain that this reptile went extinct at the same time as the dinosaurs. But then there are other scientific experts who believe that the dinosaurs did not go extinct at all. They believe that the birds of today are nothing other than the living descendants of dinosaurs.

For that matter, no one has ever come up with a suitable definition of a dinosaur. We all speak of dinosaurs, or at least I do, and we may have some vague notion of huge prehistoric monsters wandering the earth. Indeed, many of them were huge and they definitely lived many millions of years ago. It is also true that many species of dinosaur were quite small and they lived along side the huge dinosaurs. The scientists assure us that animals such as reptiles and mammals also lived at the same time as the dinosaurs, but that these species of reptile and mammal were very small and that the dinosaurs ruled the earth. Apparently our extremely distant ancestors were very small mammals which spent most of the day hiding in holes in the ground, coming out to feed and stretch their legs only after dark. With the monsters which were ruling the earth at that time, who can blame them? The trouble is that some of those huge monsters are still running around.

As best I can gather, about two hundred years ago, the scientific community was puzzled to find the bones, or at least the fossilized remains of bones, of huge animals. They searched the bible to find some reference to such animals, and came up empty handed. Of course the only reason they found nothing is because there is nothing to find. The bible makes no reference to any animals which match the description of dinosaurs. There are those who say this proves that they did not exist. They most certainly did exist, and in fact some of them still exist.

The scientists of those days were doing their best and they were truly puzzled. They faced the fact that these animals previously existed, even though the bible makes no mention of them, and they coined a word for them. They called them dinosaurs, which means terrible lizards or some such thing. The name has stuck. Although the name is popular, it is not accurate. Lizards are reptiles, and dinosaurs were not reptiles.

This brings us back to the question of just what exactly is a dinosaur. We know that they were definitely not mammals, since mammals give birth to live offspring. Any woman who has ever given birth can testify to that. Dinosaurs used to lay eggs, so the idea of them being mammals is out of the question. Reptiles also lay eggs, but it is pretty well agreed that dinosaurs were not reptiles. Within the last few years the scientists were shocked to find that dinosaurs had feathers. At least, some scientists were shocked. Other scientists were not at all surprised. Those who were not surprised were those who suspected that dinosaurs were nothing other than birds, and in fact they still are. Once again, to stress the point, birds are dinosaurs.

This is not to say that the so called dinosaurs of millions of years ago are the same animals we have with us today. To be sure, species can and do go extinct. It is also true that species evolve and eventually new and different species emerge. This is almost certainly just what happened to dinosaurs.

An example of species that very likely have a common distant ancestor is the so called big cats of today. Such animals as lions, tigers, leopards and jaguars are separate species and as such are not capable of breeding successfully. Except for the fact that they can and they do copulate in captivity and give birth to offspring.

This was not a great shock, since captivity is an artificial environment and such species normally never associate in the wild. Unless it is for purposes of killing each other, which happens on a regular basis.

Other species have been known to engage in such behaviour, especially with a little prompting from our species. Horses have been known to copulate with donkeys, for example, and in fact no great deal of encouragement is required on behalf of the human owners of such animals.

The offspring of this unnatural union is of course the animal referred to as mules. These animals are sterile, so that the scientists can still maintain that horses and donkeys are separate species, which they most certainly are. This makes the scientists very happy, since the scientific definition of a species is quite clear: members of different species are not capable of copulating and giving birth to offspring, unless of course that offspring is sterile.

The trouble is that lions and tigers, for example, while in captivity, tend to copulate and give birth to offspring which are definitely not sterile. We know this for a fact because as soon as these offspring come of age and engage in a little hanky panky, they also give birth.

Does this mean that all big cats are the same species? Of course it means nothing of the sort. I am of the opinion that the scientific definitions of species are mere guidelines, not laws engraved in stone. That which happens to animals in captivity is completely artificial, not to be confused with behaviour which happens in the wild.

Our species is no exception. The fact is that all great apes, including ourselves, have a common ancestor. This is not to say that we are descended from apes. If we were descended from apes, then why are apes still with us? It is because the apes, such as gorillas, bonobos, chimpanzees, and orangutang have the same distant ancestors that we have. That distant ancestor has long

since gone extinct, and the descendants of that ancestor have diverged to the point that our genes have no chance of mingling. We are clearly a separate species, with no chance of our genes mixing, as occasionally happens with the big cats. On the other hand, there have in the past been various other species of human, such as the Neanderthal. Is it possible that our species mingled with the Neanderthal and that we are effectively a hybrid species? There are scientists who are trying to determine that very thing. The fact is that there have in the past been various species of human which walked the earth. What is more, there is still a separate species of human walking the earth, living among us right here in North America. But more on that subject later.

For those of us with a scientific background, it just makes sense. And as long as we keep our religious beliefs separate from our scientific work, there is no conflict.

Not all examples are quite so clear cut. The grizzlies and polar bears are very closely related, just as they are to the Alaskan Brown bear. The scientists may argue over the term species and sub species, and they can fly at it. Such terms have no interest for me. I really do not care just what they call them. In the wild, under normal conditions, they tend to avoid each other, so that at least meets my home grown definition of separate species.

At least they used to avoid each other. That situation is changing due to climate change. Most people call this global warming, and in fact the climate is anything but static. This change has been going on since the climate first formed, in one form or another, so the fact that this change is taking place should come as no great shock.

This certainly has an effect on the bear population, so that grizzlie bears are now travelling further north and polar bears are now travelling further south, and nature is taking its course. Now polar bears are mating with grizzlies and I am sure the offspring are absolutely not sterile. This does not mean that the polar bears are going extinct, just that the species are evolving due to factors which include climate change. Species do that, including ourselves. When necessary, we change, we adapt. We do that which we have to do in order to survive. As for those who maintain that humans are responsible for this global warming, I would ask just what caused climate change in the past?

The point is that species evolve, and after millions of years they may form new species. And true, at some point they are not capable of successful copulation between species. It is also true that some species tend not to evolve a great deal. Scientists refer to these as very successful species, as indeed they are. Such species include reptiles such as the crocodile, and now we can add the pterodactyl to the list. In the last hundred million years these species have evolved very little.

It is only natural that we fail to recognise birds as dinosaurs, partly because they are generally depicted as huge brutes, waddling their way through an ancient swamp. We may never know exactly just what they used to look like, but it is safe to assume they resembled modern day birds. It would likely be more accurate to think of dinosaurs as very colourful, very agile rather intelligent animals which lived millions of years ago. And in fact they are still alive.

I am convinced that it is not only true that dinosaurs alive in the form of birds, but also other species, long thought to be extinct, are still alive. The pterodactyl is one of those species.

As I consider myself an intellectual giant, all athirst for knowledge, I decided to dig deeper. Since nothing else seemed to work, I asked more questions. You may have heard that old expression – when all else fails, and you are hopelessly lost, break out the road map. So they gave me direction.

I should add that we are all entitled to our beliefs. We all have our beliefs, and no one has any right to impose their beliefs on us. We should all respect each other. Live and let live, say I. They believe that the caves in these mountains lead directly to hell. They tell me that they have heard Satan open the gates of hell just after sunset, and turn loose his bird. They also say that the flapping of wings sounds like the flapping of leather.

These are the beliefs of the people with whom I live. I respect their beliefs, even though I do not share them. I do not believe that this is an evil spirit, a bird straight from hell. That is the good news, as the old joke goes. The bad news is that it is the next best thing to a demon from hell, or the next worse thing, depending on how you look at it.

It is clear that this reptile hunts strictly in darkness. And it hunts in open areas. With a wing span of a 10-12 meters, that is 30-40 feet, it has to avoid heavy timber. The elders all agree that if it catches someone outside after dark, in an open area, then God help that person. It hunts in much the same way that an eagle hunts fish. They fly in, sink their claws into that poor soul, lift that person up and carry them away. This is not at all a pleasant way to die.

Now to prove that this animal exists, we need to know a bit more about it. From the information provided we can draw certain conclusions. At least I can. Feel free to agree or disagree. Be my guest.

Since these animals emerge from caves, they must not be birds. Birds do not nest in caves. And since the flapping of the wings sounds like the flapping of leather, it means this animal does not have feathers. No feathers, no bird. This animal is a reptile; I might add that only the elders say that the flapping of the wings sounds like the flapping of leather. The young people, who have heard the flapping of the wings and survived, do not say that. When asked, they say that the flapping of the wings sounds like flapping wings. They think the question is rather stupid. Could be, but it tells me that these people were not coached. If this was all part of some sick joke, and the young people were coached to say certain things, then they may have responded that the flapping of the wings sounded the flapping or leather. And I had to be sure.

These reptiles nest in caves in mountains, but mountains which I can only describe as perpendicular mountains. By that I mean that the face of the mountain is vertical, meaning straight up and down, and the top of the mountain is horizontal, or flat. One of the elders, whom I will refer to as Grandma Ruth, sent word that when this reptile comes out of the cave, they climb to the top of the mountain. This makes sense, since if an animal that size was to jump off the mouth of the cave and open its wings, it would likely fall to the ground.

In case you were wondering, that little tidbit of information came when I was discussing the possibility of repelling down the mountain and going into the caves with a shot gun and a flash light. The elder mentioned that hand holds were likely available, but recommended against such a suicidal course of action. No doubt the animal climbs to the top of the mountain, which is flat, and uses it as a runway of sorts. I am sure it runs along, flapping its wings, and jumps off the edge. I suspect that the sound of the animal first opening its wings is the sound which people hear shortly after sun down, and I am guessing that just before day light, the animal snaps its wings shut as it flies back into the cave.

In my opinion, this animal is at its weakest when it first emerges from its cave. I am sure that a trap placed on top of the mountain would be the best way to prove its existence.

I suspect that there are also wind currents at the top of the mountain which the animal uses to assist in becoming airborne.

For some time I was wondering how the animal could climb to the top of the mountain. A little research clarified matters. I made the mistake of thinking of this animal as a bird and we do in fact refer to it as such, even though it is a reptile. This particular reptile has three fingers emerging from each wing. The scientists suspect that the animal used these claws for climbing when the animal was living here millions of years ago. No doubt this is true, and it is still true.

My friend Brian pointed out that the symbol of the devil bird, as it is called around here, is frequently carved into totem poles. He assures me that on the coast it is generally referred to as a thunder bird. Apparently most tourists confuse that carving with that of an eagle. I should add that other names for this animal include Satan bird and demon bird. Then too, Brian also says that this animal is not a bird at all. He thinks that this reptile is the Beast in the form of a bird. By that he means that it is Satan in the form of a bird. I gather there are a great many people who are of that opinion. That gives some indication of just what people think of that reptile.

As proof of the fact that the First Nation artists are carving a symbol that represents the Thunder Bird, as I believe it is called on the Coast, Brian mentions that the symbol for the tail is also carved into the wood. The tail of the pterodactyl is quite distinctive, since it ends in the shape of an arrowhead. I gather the First Nations people cannot imagine how anyone can confuse the carving of a Thunderbird, as they call it, with that of an eagle.

Chapter 2

More on the Devil Bird

In the First Nation communities the older citizens are referred to as Elders. This corresponds to a term more commonly used in non First Nation communities where they are referred to as seniors. The Elders in these communities are held in the highest respect, properly so.

There are various stories concerning this devil bird. None of these stories are terribly pleasant. The Elders rather surprised me when they sent word which indicates that this animal has more brains then I would expect from a reptile. Their experience from spring trapping, when they shoot beaver, is that the devil bird comes around night time if they have shot beaver during the previous day. Clearly, the animal has learned to associate rifle shots with food. When the boys did not shoot their rifles, the animal did not bother them during the night.

Grizzlies have also learned to associate rifle shots with food. As many hunters have learned the hard way. At least it came as a rude shock to me, but that is part of a whole different set of stories. But then grizzlies are rather intelligent mammals.

Grandma Ruth also told me about the time she was at the old village, on the banks of the Ingenika River, at a place we refer to as Grassy Bluff. That particular settlement was built at the base of a perpendicular mountain, as I call it. Of course the pterodactyl has a nest on that mountain, and frequently hunted in the villages below their nests. In fact, my friend Barbara, who lived there for many years, mentioned that the sounds that reptile made after dark just became background noise. It can make a wide variety of sounds, which resemble that of a dog barking, a woman screaming and a baby crying. There are also other sounds that the animal can make, but the people to whom I have spoken cannot seem to replicate them.

Of course, Grandma Ruth was well aware of the existence of these reptiles, and she knew better than to be outside after dark. So as she put it, after the sun was down and it was almost dark, she heard the flapping of the wings and she was not surprised. The problem was that she was outside the safety of the cabin and she knew that she was in serious trouble. Of course she immediately ran for the safety of the log cabin, and she could not see the devil bird, as she called it, but she could hear the flapping of the wings. The sound of the flapping became louder and louder and she knew that she was the target of the reptile. She also knew that the cabin was just too far away and that she was not going to make it to the cabin. In desperation, she crawled into the only shelter that was available. That shelter was a dog house, and it saved her life.

There are other stories of people who have been attacked by the pterodactyl.

Just north of our village, there is another village, about twice the size of Tsay Keh Dene, that of Kwadacha. It used to be called Fort Ware. It is about seventy five kilometres north of here, or about fifty miles. To get between the two villages, we have to travel along the gravel roads. We tend to call them logging roads, which they are, although mining trucks, lowbed trucks and of course pickups use them. Not too many cars travel on these roads because such vehicles cannot stand the punishment of these gravel roads. The roads are maintained by the Forest Service, and the roads are given names. They pretty well have to, since all vehicles which travel on those roads have to use two way radios. As a rule, vehicles heading south are considered loaded, and those heading north are called empty. Loaded vehicles are considered to have the right of way, and they are re quired to call their kilometers. The logging roads are generally rather narrow, but they do have wide spots in the road at regular intervals. The vehicles which are empty, or heading north, are required to get out of the way of the loaded vehicles, meaning the vehicles heading south. That is where the wide spots in the road come in very handy. What happens is the loaded vehicle may call a certain kilometre followed by the name of the road, such as twenty two on the Finlay main line. The kilometres are posted quite clearly, or at least they are supposed to be posted on a regular basis with very clear signs. Each main line also has a posted radio frequency, and only that particular frequency can be used on that stretch of road. It is absolutely essential that loaded vehicles call their kilometres, or their clicks as we call them, and that empty vehicles pay strict attention to their radio. It is up to the empties to get out of the way of the loaded vehicles.

The side roads off the main line lead to logging areas, which we call logging blocks. After the timber is harvested, the roads inside that block are deactivated. That just means that a machine, usually a backhoe, goes into the area and tears up the road at regular intervals. At that point the logged out area is replanted.

To get to Kwadacha, there is a road called the Russel main line. It follows the Finlay River north to the other village, only on the west side of the river. On the east side of the river, there is another road, called the Finlay main line. Traffic on the Russel main line uses a different radio frequency than traffic on the Finlay main line. In each case, the vehicles calling their clicks have to specify which road they are travelling. The empty vehicles may or may not call their clicks.

Lynn was another girl from the village who had another close call one night with the pterodactyl. She was driving from Tsay Keh Dene to Kwadacha and she had a flat tire. Of course she started to change the tire, and fortunately she had dogs with her. The dogs gave warning. They were terrified. They were trying to crawl under the vehicle or climb into the vehicle. She jumped into the vehicle and sat there all night while she heard the sounds of flapping wings. It was a long night. The only thing between her and the pterodactyl was the windshield. Just before daylight, the animal left.

What else can we say about this animal? There are only two ways of flying around in the dark. The first is by use of excellent eyesight and the other is by echo location. Bats use the echo location method which is similar to radar. But since the pterodactyl, which the locals refer to as the devil bird, makes various noises which we can hear, it does not get around using echo location. So it must have excellent eyesight.

Eyewitness sightings of this animal are rare, but they do happen. They are occasionally seen in the moon light. This is what happened to Frank, one of the boys in the village. As a teenager in the playground of the school, just across from the store, he saw a huge flying animal, about 10 feet off the ground, or three meters if you prefer. He said the wings were as wide as the store overhang or about 10 meters- 30 feet. He also mentioned that the tail was as long as he was tall. His sister was also there with her friend. As Frank pointed out the animal to the girls, his sister got a glimpse of it and the other girl did not see it but heard the flapping of the wings. All three teenagers stayed up all night hoping it would return. It did not return that night and they were disappointed. But they lived to see the sun rise the next morning.

There are some people who had a favoured camping spot around 38km on the Finlay main Line. They no longer camp there, as it is right across the river from Buffalo Head Mountain. They were all too frequently harassed by the pterodactyls, in the darkness.

An incident happened several years ago in Kwadacha, where a stallion was found dead in the morning. The carcass was found beside a tree with all the branches broken off. Also the testicles and part of the intestines were found at the top of the tree. The Elders say that without doubt it was killed by the devil bird. The Elders know exactly how predators kill and they were not surprised. No doubt the flapping of the wings broke the branches of the tree. I suspect the animal was in a feeding frenzy or as the biologists say, something triggered its predatory instinct.

This is not terribly uncommon. This animal does on occasion go after moose and elk also.

The boys also tell me that in the old village at Ingenika Point, a pterodactyl was chasing them around. Since they could not see it, it was terrifying. They responded by shooting at the sounds it was making in the darkness. In that case, they were not able to land a lucky shot.

There was a similar incident many years ago when two of the boys, who are no longer with us, were attacked by the devil bird, as they call it. Since the names of the deceased are not to be spoken after they pass on, I will not use their names. I can say that one of them was the uncle of my wife and his buddy was with him. The details are a bit vague, but it was after sundown when the boys were attacked by the pterodactyl. They responded by shooting at the sound the reptile was making. They must have scored a lucky hit, because they could hear the sounds of the animal as it was dying. They chose not to investigate, since it was almost dark and I suspect that these reptiles tend not to hunt alone. I am sure that vacating those premises was the smart thing to do, and I gather they chose not to return in the morning. In those days, such an action was extremely rare, and gives some indication of the depth of the hatred people have for that reptile.

This reptile does on occasion land on the ground and walk around. The tracks are quite distinctive, or as the people who have seen the tracks put it, they are long and skinny. These tracks are most noticeable in fresh fallen snow. My buddy Sam tells me he saw the tracks in the snow, as well as Jack. In that case, the tracks were described as long and skinny but also widely spaced. It could be that the reptile was running along, flapping its wings, and trying to get airborne.

This puzzled me, since I thought that as reptiles, they would hibernate in the winter. Clearly they do not. I have numerous reports of the pterodactyl being active in the winter in temperatures as cold as -20 Celsius.

It is also true that on occasion this reptile also walks on all fours. My friend Vivian recently was walking in the village after dark one night when she saw an animal about the size of a dog. She not unreasonably assumed that it was a dog, until she got close to it. At that point she realized her mistake. It was absolutely not a dog, but it was a slimy animal. As she was terrified, she immediately ran into the nearest house.

For those who object that this animal is huge, I can only say that when they first hatch from eggs, they do not have a wing span as wide as a house. It is only after they are fully grown that they reach that great size. Of course, if I am not mistaken, reptiles never really stop growing, but at some point their rate of growth slows right down.

Vivian was not the first person to see a pterodactyl walking around the village. My friend Randy once saw a pterodactyl walking on the bridge leading across the Finlay River to Kwadacha. He did say that in the moonlight it resembled a man walking across the bridge.

We now know that very large reptiles can and do survive in very cold environments. The leatherback turtle is one such reptile. A fully grown leatherback can weigh as much as a tonne. That particular reptile is not only surviving but actually thriving in the North Atlantic. That is a very cold place, and yet these turtles survive quite nicely through a process called gigantothermia. Although I may not understand the process, and in fact I do not care just how the process works, it does answer the question of how the pterodactyl survives in such a hostile environment.

Several years ago, some boys from town came up to the village to do some work on the store, if it matters. It so happens that they were amateur rock climbers, if that is the correct expression. They liked to climb mountains. I cannot imagine such foolishness, but to each his own.

The point is that they climbed to the top of Buffalo Head Mountain and rappelled down. That is one of the mountains which I refer to as a perpendicular mountain. I am sure they chose that mountain because it is close to the village and rather easily accessible. At that time the logging roads went close to the top of the mountain and the roads had not yet been deactivated. This is their idea of fun, if you can believe it. The mountain is riddled with caves, some occupied by mountain goats, others occupied by the pterodactyl. I have no doubt that each animal is aware of the existence of the other.

The important point is that as they rappelled down the mountain, they looked into caves and saw a great many bones. They naturally assumed that people in the village had placed the bones there. I know this for a fact because they asked the local people if they had placed those bones in the caves. Of course they had not. They did consider the question a rather strange one. People here are not like the Egyptian Pharaohs of ancient times who buried the remains of their deceased in remote caves. But then, the Egyptians had a reason for acting in such a manner. They were determined to protect the remains as best they could. They respected their dead, and did their level best to honour them. That is most commendable. People here have no reason to act in such a manner, and human remains are normally buried or cremated. That is also a sign of respect.

The question that the rock climbers asked makes sense only if the bones they saw in the caves were thought to be human remains. I suspect that is precisely the case. If the bones were thought to be those of animals, then the climbers would have no reason to ask that question, I am sure they just naturally would have assumed that another predator, such as a mountain lion, had placed the bones there. I suspect that the bones are human remains, but they were not placed there by other humans, I suspect they were placed there by the pterodactyl.

The public has got to be made aware of the existence of this reptile. With that in mind, it may help to refer to this reptile as the Terror Bird, or the Terror of the night Sky, or the Crocodile with Wings or Death in Darkness. It is all of the aforementioned, and more.

Chapter 3

The Highway of Tears

The road from Prince George to Prince Rupert is known locally as the Highway of Tears. It is called that because so many people have disappeared on that stretch of highway over the years. In fact, hitchhikers disappear on a regular basis from that road, as well as from roads in other parts of the province. It is widely assumed that there is one or more serial killers active on these highways, picking up hitchhikers and killing them. I am not one of the people who make that assumption.

The word on the street, so to speak, is that psychological profilers have been called in to give a criminal profile of the person or persons responsible for the disappearance of so many people on our highways, and not just from the Highway of Tears. They have apparently come up with a description of two very unpleasant killers, who target separate types of people. No doubt our American friends have come to the same conclusion, since the pterodactyl is very much alive there also, and it does not distinguish between Canadian and American.

There are people disappearing from other parts of the province as well, and not just from the highways. Brian assures me that he became the prime suspect in a murder investigation when a friend of his disappeared one night. Since he was apparently the last person to see that girl alive, the suspicion naturally fell on him. Murder investigations tend not to close, so it is very likely that he is still the prime suspect. Brian is of the opinion that a pterodactyl, or a devil bird as he calls it, is likely responsible for the disappearance of that girl. I rather doubt that the police are terribly impressed by that explanation.

My First Nation friends assure me that the devil bird is widespread throughout these mountains from the Arctic to Arizona, and possibly further south. They have a huge wing span and prefer to hunt in open areas. It goes without saying that a highway is an open area. And for those who say that the Highway is well travelled, I can only say that this animal hunts in much the same way that an eagle hunts fish. It flies in, sinks its claws into its prey, picks that prey up and carries it away. Such an action may only take two or three seconds. We may also point out that just as eagles are more likely to sink their claws into fish which they can easily carry away, so too the pterodactyl is more likely to attack smaller prey. Of course, they also on occasion attack large prey.

Just what size prey can a pterodactyl reasonably be expected to carry away?

An excellent question and the answer may surprise you. The fact is that the wingspan of this reptile is three or four times longer than the wing span of the largest bird which currently flies, which I believe is the condor. Does this mean that the pterodactyl can carry away an object several times heavier than an object which a condor can carry away?

Without a doubt, it can and I have been advised not to bore people with the technical details. Suffice it to say that it is quite capable of picking up an individual that weighs in the neighbourhood of one hundred to one hundred fifty pounds, or forty five to seventy kilos.

For those of you who consider this an over simplification, I say you are right. I am trying to make the point that this animal is quite capable of capturing and carrying away a rather small adult. And in fact they do just that.

The First Nation girls to whom I have spoken are of the opinion that the reason so many of the hitchhikers who have disappeared are First Nation girls is because a great many of them are from broken homes. Sadly, I suspect they are right.

I am of the opinion, which is shared by many of my First Nation friends, that most of the hitchhikers who disappear have been killed by this animal. It is also my opinion that many of the people who have disappeared have not been reported.

Of course, it is not only hitchhikers who are prey for the pterodactyl. Anyone in an open area in the darkness is potential prey. This includes people who may just decide to go for a walk. It could include people who have broken down on the highway, perhaps with a flat tire or some other mechanical problem. Then again, there are motorists who occasionally pull over to answer the call of nature. In the darkness of these mountains, that can be a very dangerous thing.

On this project I have received a great deal of information from First Nation girls. They tend to pay particular attention to the Elders, and it has been my experience that girls tend to make the best detectives. Minor details which guys tend to overlook may be carefully scrutinized by the girls. These minor details tend to be absolutely crucial. Or it could be that the girls tend to be less trusting by nature. I cannot say that I blame them; anyway, they are of the opinion that the public has got to be made aware of the existence of this predator. I am writing this largely as a result of their prompting. Further, it is their opinion that a great many First Nation girls leave home at an early age, for a number of reasons .That is the reason so many missing people are First Nation girls. Most have no money, go hitch hiking and disappear. It is our sincere hope that the awareness of the existence of this huge predator, will cause people to think twice before hitch hiking. At the very least, if outside and darkness is coming on, then by all means try to find shelter, even if only a stand of timber. There is no safety in an open area.

This is not to say that people are completely safe in heavy timber. This animal can and does on occasion land and walk around, sometimes on two legs and sometimes on four legs. It is a predator, always dangerous, never completely predictable.

I have no doubt that people are also being killed by this reptile in the Yukon, Nunavut, the Northwest

Territories and anywhere else we have these mountains. That includes Alaska and such states as Washington and points south. No doubt our American friends have come to the same conclusion that the people of this country have reached. They likely think that there is a prolific serial killer on the loose, running up and down the highways killing people. Rest assured, the pterodactyl has killed far more people than any serial killer. What is more, they are continuing to do so.

Since the pterodactyl is widespread throughout Alaska, it stands to reason that it is also across the Bering Strait in Russia. No doubt there are a few mountains in Russia. The only difference is that in Asia the pterodactyl is called the dragon. For that matter, Asia is famous for its mountains. Since without a doubt this reptile is in Russia, it is almost certainly also in other countries such as Japan, Philippines, Korea, Laos, Thailand, and Viet Nam. Not to mention China, where they have even named a year after it. Not unreasonably, they call it the year of the dragon.

For those who say that the dragon of China is a myth, I can say that the Chinese have named twelve years after twelve animals, not eleven animals and one myth. That so called myth is the same so called myth of North America; only here we tend to call it the thunder bird, the devil bird, the Satan bird or the demon bird.

Except that it is not a myth, it is a superb predator, it is the terror of the night sky, it is a man eater, and it is a pterodactyl.

Chapter 4

The Takla Lake Monster

My friends, Roy and Kathy, assure me that there is a second huge predator, located in Takla Lake, among other lakes in the province. The Elders around Takla say that the animal goes into caves around the lake. It must be huge. The story is that a couple of the boys shot a big bull moose in August. They went across the lake in a boat, shot the moose and decided to tow it across the lake with their boat and motor. That time of year, the bulls are at their finest, so in August we always go after the biggest bulls we can find. It is only after they go into the rut that they quit eating, start fighting and lose all their fat. They also do that which comes natural with the cow moose. By the time the rut, as it is called, is well under way, the bulls have lost all their fat and what is more, the meat stinks. I tend to favour the expression that the cartoon characters in the Disney movie Bambi use for the mating season. They call it being twitterpated.

No doubt the bull moose was very large and very heavy, since large and heavy tend to go together. Likely they thought to drag the moose to the other side and let the girls butcher the animal. They had a boat and motor so why not use it? And who better to do the hard work of butchering than the wives? And rest assured, those girls know how to butcher a moose. A fine plan, but it did not work out quite the way they had in mind

Part way across the lake, something grabbed that huge moose and took it under water. Worse, it was dragging the boat backwards at a rather alarming rate of speed. The boys in the boat did the smart thing and cut the rope. As a result, they did not get wet. They did see the water swirling around and never saw that moose again.

That was not the first sighting of the monster in Takla Lake. Even though the First Nation people there are aware of it, they have apparently not given it a name. Or at least, not a name of which I am aware. There are other names which apply to it, such as ogopogo. It sounds like a name for a cartoon character, which this animal is certainly not. It is in fact another reptile, and another man eater. It is a plesiosaur. It is another huge prehistoric reptile from the age of dinosaurs that supposedly went extinct along with the dinosaurs. The plesiosaur absolutely did not go extinct. It is another predator that is superbly adapted to its environment, very likely not changed from its appearance of millions of years ago. That makes it supremely dangerous, just as the pterodactyl is extremely dangerous. I suspect that its main prey is fish, but I am sure that it will eat any prey that comes into the water.

Several years ago, there were a couple professional scuba divers doing work for the government in Takla Lake. The word is that those boys came out of that water a lot faster then they went into it, and furthermore, they had no intention of returning to that water. They said that there was something in that water, and it was clearly huge and fearsome.

I gather that the language they used left no room for any misunderstanding.

This did not stop the younger people from swimming in Takla Lake. The story is that one girl was on a tire tube in the lake and saw something that terrified her. She too vacated the Lake in record time. From the description, it resembles a crocodile. I gather this animal is also in other big lakes. I suspect it may be the same animal in Europe that they call the Loch Ness monster.

I suspect that the simplest way to prove the existence of this lake monster is to do the same thing those boys did. A wild animal is a hungry animal, so a simple act of trolling should be able to prove their existence. There is no need to use a big bull moose. I suspect that a moose quarter or a deer would suffice. That and a proper hook, a strong line and a big boat and motor should get results.

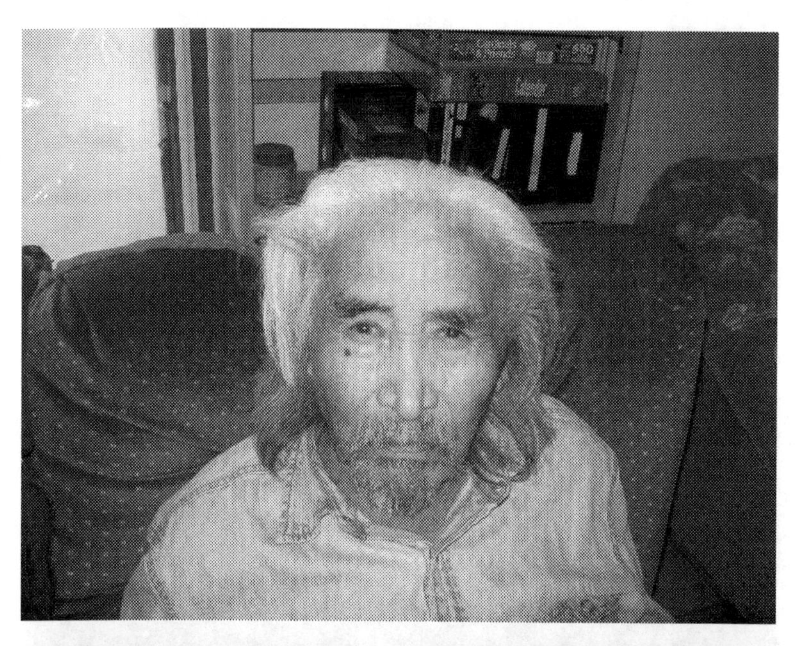

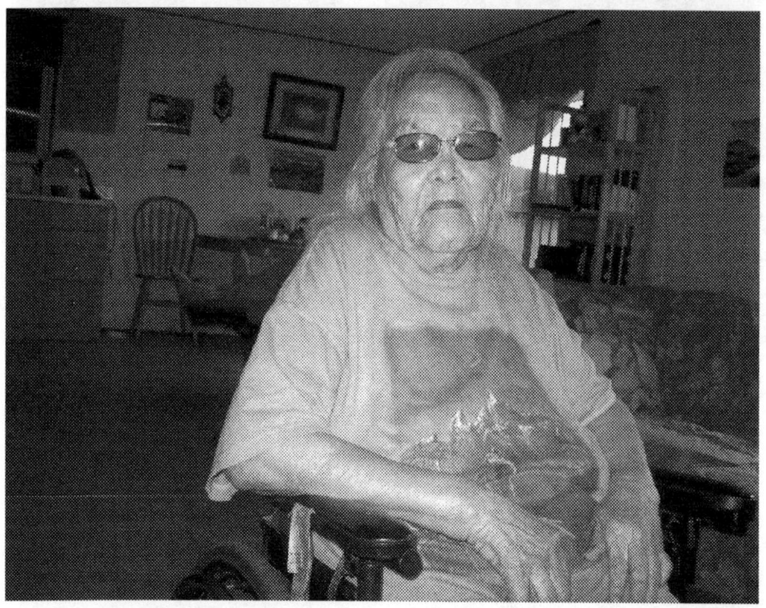

Elders of Tsay Keh Dene – Isadore and Julia Izony -who advise the author to leave the giants alone.

Elder, Mike Abou from Kwadacha, says the mammoth go into caves for the winter and that the giants go south for the winter.

This cabin is the place that the author met the giant many years ago. At that time the porch was open at the front of the cabin. The giant would come onto the porch after dark. The author thought that it was a bear at the time. The window at the side of the cabin was the place where the author was standing when the giant dropped by just before full dark. The giant was shaking nearby trees.

Buffalo Head Mountain, an example of a perpendicular mountain, where the devil bird has a nest.

The view from Hydro Creek, looking down the lake, at the point where the three young people saw a Giant drinking water.

Rita McIsaac, the wife of the author.

The store at Tsay Keh Dene. A devil bird was seen in the moonlight in front of the store. The wingspan was reported to be as wide as the store overhang.

Chapter 5

The Hairy Elephant

My mother in law, who has since passed away, grew up in this country at a time when it was extremely remote. She lived her whole life in these mountains, and in those days there were of course no schools. She spoke broken English, and never even set foot in town until rather late in life. That event was an absolute culture shock.

I was just as shocked I am sure, when my wife told me that her mother had recognized a picture of an elephant. That was the day I realized that I was missing something really big. I mean Really Big. I am talking about a woolly mammoth.

It takes a great deal to scare the elders. I admit, I did not think it was possible. That was before that told me about the Devil Bird and the Hairy Elephant. The woolly mammoth scares them. The woolly mammoth is not a pleasant animal. I am sure it is a vegetarian, but it appears to have a strong dislike for people.

There are some scientists who think that intelligent animals are capable of such emotion as hatred. If that is the case, and I believe it is, then the woolly mammoth, as an intelligent animal, hates us.

I suspect there is a good reason for this hatred. My First Nation friends tell me that the mammoth was alive in the Midwest in the 18th century. They also tell me that it was alive in Albert in the 19th century. And it was definitely alive in this province in the 20th century.

If I am right, at the time of the European Invasion the mammoth was wide spread across North America. As settlers pushed west, they pushed the mammoth ahead of them. Completely understandable, since those people all planted crops and no farmer wants a ten ton vegetarian in his garden.

Farmers in Africa are familiar with the problem and I am sure some of them would love nothing better than to shoot every elephant that comes onto their farm. They are restricted by laws. The farmers of North America of many years ago were not hampered by those laws. I am sure they shot the mammoth at every opportunity. It is just possible that the mountains of Western North America contain the remnants of a once great herd of mammoth. Not convinced? Consider the facts.

The Elders assure me that if the mammoth catches your scent, it is likely to run you down and kill you. Your only defence is to run for the nearest swamp. That animal is very heavy and cannot go into swamps, as it will just break through. In winter, it is not an issue, since those animals go into caves.

My wife tells me that her grandfather once found the remains of a mammoth. The animal had used its trunk to push together trees to form a house and there were a lot of mammoth droppings in the area. Of course the scavengers had been busy at the remains. No doubt the animal was sick and had constructed a shelter for itself. After some time it had died. I am guessing this happened in the early 20th century.

More recently in the spring of 1965, two of the boys from Kwadacha were on the trap-line, trapping beaver. One of them saw a mammoth and immediately ran back to his buddy and both ran for their lives. Their weapons were 30-30's and figured there was no way a 30-30 bullet would bring down a mammoth.

Those two individuals are no longer with us but they told some lads who told me that story. There are other stories of mammoth terrorizing people, and it is possible that there have been more recent sightings. I am currently trying to track down some people who have reported seeing animals which match that description.

Chapter 6

The Rubber Faced Bear and the Wilderness Wolf

In 1958 when Archie was just a teenager, he and a few of the boys found sign of a huge bear. It was clearly a monster of a bear. All bears mark their territory by standing on their hind legs, reaching up as high as they can and scratching a tree. This serves as a warning to other bears to stay away and probably keeps fights to a minimum. After all, there is no point in picking a fight with an adversary who is far bigger than you are.

In this case, the boys were amazed to see that the claw marks were so high on the tree. They are accustomed to seeing big bears, but this one was clearly record size. The tree was a huge pine and yet the bear was so big and strong that by scratching and pushing on the tree, part of the roots were exposed.

This was 50 years ago, and Archie, now an elder, was just a lad. Sometime after that the boys shot a huge Rubber Faced bear. If took 15 shots from a 30-30 rifles to bring it down. They called it the Rubber Faced Bear because it has no hair on its face or on top of its head. It also has a short nose.

The Elders think that while in caves for the winter, the bear rubs its face with its paws and all the hair comes out. I am of a different opinion.

One scientific theory is that climate change in the form of Global Warming caused the extinction of several species of Mega Fauna at the end of the last ice age about 10,000 years ago. In case you were wondering Mega Fauna just means Big Animals.

Well, in the last 100,000 years we have had three ice ages. That means that we have had three episodes of climate change in the form of global cooling and three episodes of climate change in the form of global warming. If global warming caused the extinction of these species then what happened the first time? The fact that the so called Mega Fauna were alive at the end of the last ice age proves that they survived five previous episodes of climate change, three times in the form of global cooling, and twice in the form or global warming. I am of the opinion that climate change did not cause the extinction of these mega fauna, mainly because at least some of them are not extinct.

I suppose by now you are wondering what species supposedly went extinct. I am so glad that you asked. The theory is that climate change caused the extinction of the Woolly mammoth, the Cave Bear, the Dire Wolf, the Sabre Toothed Cat and the Giant Sloth.

The Cave Bear, which is also known as the Short Faced Bear or the Mega Bear, matches the description of the Rubber Faced Bear. There have been other more recent sightings. This cave bear may well be the world's largest bear. A full grown cave bear may stand six feet at the shoulder, or one point eight meters.

In case you are wondering just why there are so many names for this one species of bear, consider the mountain lion. There is a scientific name for that animal, but it is also referred to by numerous other names such as cougar, puma, and catamount. Most scientific names tend to be rather non pronounceable.

The Rubber Faced Bear is also called the beaver killing bear, by First Nations people. It is possible that the wide spread killing of beaver, due to the fur trade, reduced their numbers. With less prey, it is possible the population of bears which preyed on those beaver was reduced. Now that far fewer beavers are being killed, perhaps the bears are making a come back. I should add that the First Nation people to whom I have spoken have no knowledge of the sabre toothed cat or the giant sloth. I suspect that those animals are extinct, but I have no idea just what caused their extinction.

This brings us to the Wilderness Wolf, almost certainly the Dire Wolf. The world's largest canine. The First Nation people are well aware of this animal, since it comes into the village after dark and kills dogs. Several people have spotted that animal. They tell me that it is even taller at the shoulder than a deer.

Chapter 7

The Giants Living Among Us

In 1972 or thereabouts, long before there were any roads in this country, a 13 year old girl disappeared from Kwadacha, or Fort Ware as it was then called. The RCMP were called in, a search party was organized and for days people looked for that girl. After several days, the search was called off. She was given up for dead. No child can survive for very long in the mountains. After 10 days, she walked into the camp of her relatives. She was very much alive and quite healthy. So what happened?

The fact of the matter is that Giants live among us here in the mountains and forest of Western North America. Most of us have heard of them. The names vary from location to location, but include Bigfoot,

Yeti, and Abominable snowman. The Elders know them well, and refer to them as Giants or Ice people or Stink people.

It is significant that the Elders refer to the Giants as people. The fact is that these Giants are human. They are just a different species, but human none the less. The Elders have a great deal to say about the Giants. First and foremost: Leave them alone. This is an excellent piece of advice. They respect us and leave us alone and we should show respect for them and not invade their privacy.

The Giants are clearly nomads, coming north for the summer, and then going south in the fall. Fully grown, they may average 8 feet tall-2.5 meters, although some are closer to 9 feet tall-2.8 meters. They may weigh in the neighbour hood of 1,000 lbs, 500kg. We can safely say that they are the Worlds largest ape, since they are twice the size of gorillas. In fact, they are the largest apes ever to walk the earth,

What else can we say? A great deal, since this is a free country. They are hi-pedal, which means that they walk on two legs. They are also top predator in North America, since all other predators, including grizzlies, run from them. Predators that do not run from the Giants can expect a dramatically shorter life span.

The Giants tend to be gentle, and have never been know to hurt a child. As big as they are, they move as ghosts through the forest. They are right at home in the forest.

They tend to be active at night, probably because they know how violent we can be.

My friend Allan took a shot at one recently. He was in an isolated area, and Allen thought it was a bear. That was a mistake. That giant was kneeling down and drinking water and when my buddy fired, the Giant jumped up and looked him in the eye and ran off into the forest, on two legs of course. My friend was terrified and ran to the nearest camp, where his buddies were. His friends offered to take rifles and go back to have a look. The offer was rather wisely refused and to this day he hopes and prays that the Giant does not hold a grudge. After all it was an honest mistake and we all hope the Giant understands that.

I should add that the Giant ran off into the woods and did something that is rather characteristic in these sorts of encounters, even though it is rarely reported. Once the Giant was out of sight he shook a tree. Such an action is very likely their way of giving us warning, letting us know that at least in this case, they certainly did not approve of that sort of behaviour. By that I mean that he did not approve of being shot at. Not that anyone can blame him for that. It is pretty well taken for granted that any misguided soul who kills or injures a Giant will almost certainly not get out of that forest alive. The Giants take care of each other and if anyone kills or injures a Giant then their friends take action. The Giants also dispose of their deceased, likely burying them.

There are some people in a different village, those who found a sick and dying Giant and cared for that Giant as best as they could. This gives an indication of the respect the First Nations people have for these Giants. They cared for that Giant because of their concern. The Giant died, despite their finest efforts, and in the middle of the night the body disappeared.

No doubt other Giants picked up the body and carried it away. I suspect they bury their dead in caves.

There are indications that the Giants spend a good deal of time in caves. It would make sense, since the caves would provide shelter and warmth, not to mention a place to hide from us.

In the case of the girl who disappeared from the village for several days, the Elders in the village were not terribly surprised that a female Giant had borrowed her. They figure that the old girl had lost her child and was lonely. There was no malice intended in grabbing that child and for 10 days the Giant provided for her, feeding her berries and raw meat. Then she was brought back close to the village where her relatives were camped out. She walked into that camp dirty but healthy.

The people who have seen the Giants close up all agree that the faces are black and shiny. The girls also tend to use such adjectives as beautiful and gorgeous. The boys who have seen them up close tend not to use such adjectives.

The experience of Dorothy, a non First Nation girl is rather typical. As a teenaged girl at a resort in the southern part of the province, she was told that under no circumstances could she leave the compound and climb the mountain next to it. The mountain had recently been logged off and the road to the top had switch backs, as we call them. That just means that the road zigzagged up the mountain. The usual vegetation grew back immediately after it was logged, which includes berries and willows and such. The vegetation was over waist high.

Is there anything sweeter than forbidden fruit? In no time Dorothy and her friend were hiking up the logging road to the top of the mountain. Sure enough they found trees and rocks, very similar to the trees and rocks at the bottom of the mountain. They were not impressed. They could not understand what was so special about the mountain. What was so important that they were not allowed up there? They soon found out.

As they were walking down the mountain, sticking to the logging roads of course, they approached a switchback. The Giant was just off the road, crouched down on all fours, waiting for them. Just when the girls came to the curve in the road, he jumped up right in front of them. The girls took a short cut down that mountain. They went straight down, through the brush and willow, screaming at the top of their lungs.

Of course, from the time the girls left the resort and started up the mountain, their every move was being watched. The Giant who surprised them was likely an adolescent, a teenager like themselves. This is the Giant way of saying boo. And rest assured they do that. They do have a sense of humour, which not everyone appreciates. Of course the people at the resort were not the slightest bit surprised. In fact it was pretty well just what they had expected. Dorothy is one of the girls who used the adjective beautiful to describe that Giant, although I doubt they had any more dates.

It is clear that these people are rather easily offended and when we disrespect them, they know how to disrespect us. So for those who comment on the terrible smell of these people, I say you are mistaken. I will try to phrase this as politely as I can. In response to an insult, the Giants let loose a blast of foul smelling air from their rectal orifice. This is important, so even though it sounds crude, it must be said. The fact is that when the Giants are insulted, they fart. I know that sounds crude, and it is, but as I say, I cannot think of any polite way to say it. But then again there likely is no polite way of saying this. If nothing else it is an indication of their intelligence, their humanity. Just think: Among our species, among all our different cultures and languages around the world, there is one insult which is universal. There is no greater insult than to fart in someone's face. It is the supreme insult and they use it on us. In fact, they use that insult on a regular basis. I have spoken to a great many men who have had close encounters with those people and most men have commented on the terrible smell of those people. Most girls, on the other hand, who have had such encounters, do not mention any foul smell; apparently, men are more likely to offend those people than women. This does not surprise me terribly.

If nothing else, it gives us an indication of their humanity. 0f course it also lets us know just what they think of us. We have no reason to feel flattered.

CHAPTER 8

Bears or Giants?

I had my own experience with the giants. This happened years ago when I was living alone in the old village at Ingenika Point. We had no electricity, no roads and no running water. I was living in a log cabin about a fifteen minutes walk from my nearest neighbour. In the spring of that year, I thought I was going crazy. I could not shake the feeling that I was being watched. That made no sense since I knew that no one was watching me. The fact is that everyone in the village knew exactly what I was doing. I just figured that this insanity was a natural result of living alone. It is not. I was being watched. I should add that previously a domestic cat had dug a hole under the cabin and crawled in there, a safe place to give birth to her kittens. The hole was not very big, rest assured. I also had a fair sized dog living with me, not a terribly good natured animal. In fact he was the meanest dog I could find.

Up North here, we have long days and short nights, in the spring. One day the sun was down and it was not quite full dark. I heard a noise outside the cabin. I was puzzled, since there was only one trail leading to the cabin easily visible from the window and I had not seen anyone approaching. Also the dog was not barking. So I went outside and sure enough, something was making quite a racket, but it was just out of sight. It was the time of night just before fall dark. I could see about five yards and this animal was at about six yards. I could not get a glimpse of it. I did get a good look at that dog, and I was not at all impressed. It was terrified! In fact, it was so scared, it was trying to crawl into that small hole the cat had made, in order to get under the cabin. I thought that dog was a coward, wrong again. That is a typical response of dogs in the presence of Giants. Dogs do not bark. Dogs do not growl. And dogs do not whine. Dogs do not run away either, which is rather strange. It can also be rather rough on dogs too, if as I suspect Giants see them as an easy meal. Anyway, I just thought it was a bear, since it was making so much noise and scaring the dog so badly. I called the dog to my side and for those of you who say that dogs do not have any facial expression; I can only say that you have not seen a dog in the presence of a Giant. That was one terrified animal. Of course he had better eye sight then I did. If I had seen that which the dog had seen, I too would have joined him trying to crawl into that little hole.

I was simply puzzled that this bear was acting to strangely. It sounded like it was walking through clumps of willow and small trees and bears only do that when they have to. Since I had the place cleaned up nicely just like a park there was no need for any animal to walk through willow. The area around the cabin was nice and open so why was the bear walking through the willows? Then too, why was it here at all? There was no garbage, no carcass; there was no reason for bears to come around. It just made no sense. I decided not to worry about it. Just wait until morning and shoot that strange acting bear in the day light. There was no way that I was going to grab a rifle and go after that bear this close to dark. This was my idea of being smart, so I went to bed. Shortly after that it was full dark and I could hear that bear on the porch. I was determined to shoot that bear in the morning. I was not worried, because my rifles were loaded and ready, hanging on the wall just over the bed. Besides, bears always enter a cabin through a window that is closest to the ground. That was the window on the side of the cabin, not the small window higher up in front. I went to sleep.

In the morning I got up, grabbed my rifle and went bear hunting. I first checked the porch, since bears always leave a sign. That is what we call scratch marks from the claws and dropping and such. The trouble was, there was no sign on the porch. I was completely puzzled, but all around the porch is hard packed ground. There was no way a bear could walk across that ground and not leave a sign but this bear managed. Then I checked the clumps of willows and small trees that the bear had been walking through. There were no broken branches.

Of course it was not a bear. It was a Giant and the fact is that the noise I heard was that of the Giant shaking trees and willows. That puzzled me for awhile- so okay, I am easily puzzled- but then I realized that when they want to make contact with us, they shake a tree. I suppose the alternative is to walk up to the campfire and introduce themselves, but they know the response they get. They terrify us.

I suppose the polite thing is to put out food for them, just to let them know that we are friendly and want to get acquainted. But my mind at the time was elsewhere, and becoming buddies with a Giant was not one of my goals.

This went on for a couple of months, until I decided to move elsewhere. It finally occurred to me that the stories I had heard –we call them whoppers –were likely based on fact .Stories of guys being picked up by husband hunting Giants.

One of the girls in the village spelled it out for me in terms which even I could understand. She called me a dummy, and pointed out that I was on the shelf. That is a term they use for single people. And apparently even Giant girls go husband hunting. Much as I admired that Giants taste in men, I chose to pass her up. Mind you, it does give a whole new meaning to the term mixed marriage.

The fact is that truth is stranger than fiction. I was reminded that a Mountain Man by the name of Colter was a famous story teller. That boy could tell a whopper. That is the name that people use for the completely unbelievable stories that Mountain Men tell. Mind you, Mountain Men tend to call them by a more accurate term, even if it is less polite. We call them lies.

There is no harm in this. It is entertainment. It seems to come natural from years of living alone.

Colter was by all accounts a typical Mountain Man. Likely not real long on the social graces. He preferred the mountains and solitude to big city life. I can understand that.

Then one day, after being in the mountains for a lengthy time, Colter out did himself. He came out with a whopper that was an absolute masterpiece. People loved it. It was detailed, it was graphic, and he swore it was true. The more he swore it was true, the more it was loved. It never occurred to anyone that his greatest whopper was just the Gods honest truth.

This was one story that people could not hear often enough. Colter swore that there was a place where water never froze. The way he told it, even in the middle of winter, the water remained so warm you could take a bath in it, if you were the sort of soul who indulged in such foolishness. Most Mountain Men are not terribly concerned with personal hygiene. Colter was no exception.

He also mentioned that there was a place where water came squirting out of the ground as regular as clockwork, day and night, summer and winter. You could set your timepiece by it, if you were the sort of pilgrim to pack around such a useless contraption. Mountain Men have no use for such devices. They do not need a clock to tell them it is time to eat. They eat when they are hungry, just as they sleep when they are tired. Any fool knows that it gets dark when the sun goes down, which makes it a good time to sleep. Of course, the sun comes up again in the morning, which makes it a good time to get up.

Of course, Colter was the first Mountain Man to set foot on Yellowstone. His description was detailed and accurate, and completely unbelievable. It was also absolutely true. I was reminded of this when I thought I was going crazy at the cabin. I recalled the other mountain man stories of being picked up and carried away by husband hunting Sasquatch. I always thought those stories were just stories, not to be taken too seriously, just a form of entertainment. But after about two months of having company on my porch every night, I was not at all sure that it was just entertainment. I was certainly not the slightest bit amused. And I knew that I was being watched all day long. I could think of no other reason that a Giant would be sticking so close to me. So I did the smart thing. I moved away from that place. In case you were wondering, even though I was very close to them on numerous occasions, I have never smelled them. Apparently I have never insulted them.

Many people have asked me what happened to that dog, the one I incorrectly thought was a coward. I admit that I honestly do not know. It was many years ago, and since I have little use for cowards, I may have shot him. Rather tough on the dog, but then he had to die sometime. Anyway, no one is perfect.

Then again, he may have provided a meal for the Giant. I personally am not partial to dog meat, but to each his own. Who am I to judge? It could be that dog did not run due to loyalty. Well, everything has a price, even loyalty. That fool should have run.

Then again, if he had run away, I likely would have thought he was disloyal and shot him. Around here we call that being caught between a rock and a hard place. Well, he was just an animal, and he had to die sometime.

Chapter 9

More on Giants

For those of you who are of a scientific disposition and I suppose there are a few of you misguided souls, just as I am, I will point out that the technical term for these people is Gigantopithecus. Do not ask me how to pronounce it, because I do not know. Just take your best shot. I believe it is of the order of Sivapithecus, another mouthful of a word, so that their closest living relative is the orangutang. As if you care.

The point is that they are very distantly related to us, so that the chances of our genes mixing are practically nonexistent. There is another scientific theory to the effect that a separate species of human, in Europe, called Neanderthal, may have interbred with our species. It is entirely possible that the Neanderthals may not have gone extinct exactly. According to this theory, the Neanderthals merely bred with our species, and we are a mixture of Homo sapiens and Neanderthal. This theory has yet to be proven.

In the case of the Giants, they are definitely not extinct, far from it, and I suspect they may be just as intelligent as we are. They have managed to avoid us quite nicely, and most members of our species are unaware of the existence of these people. They on the other hand are supremely well aware of our existence.

From all accounts, their tools are rather primitive, at least compared to our tools. These tools may not accurately reflect their intelligence as much as their incredible strength. I suspect they have developed the tools they need to survive, just as our species has developed the tools we need to survive. The only difference may be that we need far more sophisticated tools to survive, because our species is so puny, at least compared to the Giants. And rest assured, the bow and arrow is a very sophisticated tool. I am not even sure the Giants use knives, probably because they do not need them. With their great strength, they are quite capable of butchering an animal with their bare hands.

It is just a matter of time, and very likely a very short time before we have a serious run in with the

Giants. Certainly their behaviour is changing. It is my impression that the Giants are under extreme pressure and are turning to alcohol. They are prepared to go to considerable length to get that alcohol. The fact is that during hunting season, hunters tend to go into those forests with something more than their rifles. Of course I am referring to alcohol. Without doubt, one of those Giants is going to get his hands on a bottle and get drunk. God help anyone who gets close to a drunken Giant.

Recently in Kwadacha some young people were partying by the airport. They were in pickups and it was getting dark .Then they noticed a Giant coming closer and closer to them. They fired up those vehicles and got out of there in a rush. Clearly the Giant could not resist the lure of alcohol.

Although most people do not bother, there is in fact nothing easier than making alcohol. The only thing required is sugar and water, although most people prefer adding some kind of fruit or vegetable for flavour, and the sugar in the fruit or vegetable adds to the alcohol content. It also provides a certain flavour, which at times can be a mixed blessing. By that I mean that the taste of that brew can be rather vile. At least I do not think that anything more than sugar and water is required. Either way, I do not care. The point is that a little yeast gets the process off to a roaring start. Then it is just a matter of keeping the mixture warm for a few days and the resulting fluid can be quite potent. Around here, we call it home brew, or just plain brew. It is just one more name for alcohol and while not as potent as moon shine, which is made by a different process, it can still get one very drunk. I know personally a couple people who each drank about two litres of a batch and each managed to get so drunk that both blacked out. That is the name we give to being so drunk that we cannot remember what happened. It was clearly the best batch of brew they ever made.

Until very recently some people here used to set home brew in isolated cabins and leave it for a few days. Not any more. By the time people came back, the brew was gone. At first they were puzzled, thinking perhaps that someone had stolen their brew, but then they noticed the hair that the Giants had inadvertently left behind. Without a doubt, the Giants know about alcohol and they love the stuff. They are truly human. They love the same things we love. Sadly, that includes alcohol. Not that I blame them in the slightest.

Recently three young people in the village saw a Giant. It was summer, close to dark, and the three were on quads. The Giant was thirsty and went to the lake to get some water. The lake is a man made lake, and the water is used to generate electricity, so that the level of the water fluctuates rather wildly. This particular time the lake was quite low, with a rather long steep drop. The Giant faced the problem and quite simply pushed over a fair sized tree and used that to slide down to the lake. This gives some indication of the strength of those people,

Anyway, the Giant was down on one knee, scooping up water with one hand, drinking. The young people sat there for five or ten minutes, by their estimate, watching him. He must have known they were there, especially since they sounded the horn for him. The Giant was an Elder, judging by the grey head and grey torso. After he was through drinking, he stood up and walked away down the lake shore.

To be sure, the next day the boys from the village went to that particular spot and looked for tracks. They found nothing, which did not surprise me. The Giants know how to hide

their tracks, literally.

The fact that the Giant was not at all concerned that the kids were watching him is an indication of the change of attitude that those people. I suspect they are resigned to a confrontation with our species.

The disappearance of the girl from Kwadacha was not an isolated incident. There was a similar event which happened at Tucha Lake several years ago. A 5 year old girl went missing. Of course the boys went looking for her. My friend Adam heard the girl. She was on top of a hill, and he could not see her through the trees, but my friend could hear her. She was telling someone to go away, to leave her alone. My friend had a 30-30, a lever action rifle, and chambered a cartridge into the chamber. He went up that hill and found that girl alone. When he asked her who she had been talking to, she responded –Spiderman. That girl was missing for several hours.

In another incident years ago, some of the boys shot a moose at Akie swamp. It is a preferred hunting ground, a fine place to shoot moose. It is the local equivalent of a supermarket for local hunters.

As happens on a regular basis, a moose was shot there one day, close to dark. The boys just had time to butcher the moose, as it was getting dark, and they decided to come back in the morning. This was something they had done on numerous occasions in the past. And just as they had done countless times before, they returned in the morning with some dogs to pack back the meat.

On that particular day, things did not work out quite as planned. In the morning, the moose was gone. Every little bit of it. Nothing was left. Even the hide was missing.

What likely happened is a Giant had happened along and decided to help himself. Clearly not all Giants are the shy, bashful sorts. This one saw an opportunity, and seized it. No doubt we have meddled in there food caches, so turn about is fair play. One good turn deserves another. I do not blame him in the slightest. But then, it was not my moose. It is easy to be generous with that which does not belong to us.

The fact that the Giant was able to pick up a moose, put all the meat into the hide and pack it away is another indication of the strength of these people.

There are other indications of their humanity. The girls were picking berries recently, when one of them found something strange .There in a clearing someone had placed a great many branches of berries. Huckleberries, if it matters. They were piled very neatly, about knee high. She knew better than to touch them. The clearing was facing south and was sure to get a great deal of sunlight. It is one of the ways that Giants preserve their food.

It is my impression that the Giants are respected and admired by the people to whom I have spoken. Most consider it good luck to see a Giant. For sure, as long as Giants are close to you, you need not worry about predators. It makes sense to be nice to Giants, to feed them. We should let them know that we are friendly.

Chapter 10

What now?

The information that I have gathered can be used to help gather proof of the existence of these huge species. At the risk of being accused of being redundant, I will say again that with the cooperation of the First Nation people, it should not be difficult for us, acting as citizens, to prove that they, the First Nation people, are absolutely correct.

Concerning the Giants, without doubt we have to make contact with these people. The best way to do that is to do the same thing that I did years ago, without meaning to do it. We should set up camp in a private area and wait for them to come to us. Their usual greeting is by shaking a tree. We can respond by putting out food for them. Let them know we are friendly. It will take a while to gain their trust.

I would like to see laws passed to protect these people. They should feel free to be seen in daylight, and live their lives as they see fit. All other species are protected by law. Can we do any less for other humans?

The fact is that they are different from us, and they scare us. We are intimidated by their size. Their strength is awesome. They are huge and they are hairy and they do not wear clothes. It is only natural that we fear that which we do not understand, but we are going to have to get over it. Those people deserve better than that.

I am aware that there are people who are making a career of tracking down these people, the Giants. I am also aware that although their intentions are fine, the result is that they are making the lives of these Giants very miserable. These people are human and they have the right to their privacy, just as we have the right to our privacy. They respect us and we should respect them.

I have the greatest respect for the Elders, just as I have respect for the scientific method.

The scientific community is blissfully unaware of the existence of these huge species.

Clearly the scientific method is not being followed. By that I mean that the scientific theories are not being questioned. A theory is just that, a theory. These theories are not facts, even though they are being presented as such. Such theories should be challenged, and not memorized as a child is taught to memorise the alphabet.

The fact is that there are six huge species, thought to be long extinct, which are right here in front of us. When I say huge, I mean huge. Each one is, I suspect, world record breaking in terms of size. Without a doubt, the pterodactyl is the largest flying animal in the world. I suspect the plesiosaur is the world's largest reptile. The cave bear is likely the largest bear in the world. The dire wolf is the largest canine in the world. The woolly mammoth is very likely the largest land dwelling animal in the world. Last but not least, Gigantopithecus is absolutely and without doubt the largest ape in the world and a separate species of human. To think that the scientists of the twenty first century have missed a human species right here in our own back yard is shameful. We can expect no cooperation from the scientists because they are more concerned with their reputations than they are with determining the truth. They are not about to rock the boat.

I find it strange to say the least and annoying to say the most, when people ask me about unidentified flying objects. This subject almost always arises when the topic of Giants is being discussed. The fact is that these objects are called unidentified for a reason. They have yet to be identified. As yet the scientific community can offer no explanation for them, aside from the occasional denial of their existence. Each and every one of us can choose to believe or disbelieve any explanation for those objects. I prefer to deal with facts and the facts are that a suitable scientific explanation has yet to be offered. I see absolutely no reason to speculate, especially since millions of others are doing just that and accomplishing absolutely nothing. Another fact is that there are several huge species of animals right here in North America and the act of proving their existence would accomplish a great deal. That includes the Giants and just how anyone can speak of Giants and flying objects in the same breath is beyond me. So for those people who are focused on flying objects which have yet to be identified, I can only suggest that you focus your time and energy on more practical matters.

The scientists have got to start paying attention to the Elders, and work with the First Nation Youth in order to properly manage our wildlife.

It is essential that we work together to prove the existence of these species. It should not be difficult, as long as we have the cooperation of the First Nation people. It is my experience that people tend to respond quite well to respect. Those who receive respect tend to give respect in return. Then it is a matter of passing laws to preserve and protect our wildlife, which should be very difficult, if I know politicians. I do not include the Giants in the reference to wildlife. They are human and should be given all the protection that the law gives all humans. That should be even more difficult to get those laws passed, but I can only hope that the politicians surprise me.

On the other hand, if the public is made aware of the existence of these neighbours of ours, I suspect people will show respect for them, with or without the laws to protect them. No doubt the Giants will respond to the easing of pressure the same way we respond. I am sure they will relax and become more visible, especially during the daytime, and our separate human species can make contact. I am looking forward to that day.

For our part, we have got to quit thinking of the Giants as animals and face the fact that they are human. We need to show them the respect they deserve. It is not polite to take a picture of someone, for example, without asking their permission. We should show the Giants the same consideration, and treat them as we would have them treat us. We should not gawk at them as we would an animal in a zoo. As the pressure eases, I am sure they will approach us. It is better that we let them come to us, and not force ourselves upon them. And for their part, we can only hope that they will show us a little respect and quit farting in our faces.

There is a purpose for all these different species, just as there is a purpose for all forms of life on this planet. That includes the pterodactyl, just as it includes the mosquito. One species is about as adorable as the other. Much as we may hate mosquitoes, for example, they are absolutely essential for life on earth, especially in the north.

Each spring we have a huge migration of many millions of birds. They have a rather short time to build a nest and raise their young before heading south in the fall. That is where mosquitoes come in very handy. True, the birds do eat a great deal of vegetation, but they also eat a great many mosquitoes, among other insects. Bats also eat mosquitoes, as do fish, frogs, toads and other animals. And I suspect that mosquitoes do their part to pollinate vegetation.

As a huge predator, the pterodactyl is also a huge scavenger. The people who argue over predators versus scavengers are missing the point that it is two sides of the same coin. The African lion is top predator in its neighbourhood, and it is accordingly top scavenger. Whenever possible, it steals prey from lesser predators. It hunts only when absolutely necessary. It is far easier to steal a kill than to make a kill. No doubt the pterodactyl acts in a similar manner, which nicely explains the reason this reptile would show up after dark on days when the boys shot beaver. So this completely disagreeable reptile helps to keep the country clean by devouring rotten carcasses, among other things. This sort of behaviour does not make them more lovable, but it does help to explain their purpose for existence.

To put things in perspective, bear in mind that most of us do not kill our own meals. We tend to let someone else do that for us, and we buy the product. This is a very civilized form of scavenging, and very effective. As top predators, we can do this sort of thing.

As I write this I realize that I am relying rather heavily on input from the girls in the villages. Not that there is a great deal of choice in the matter. Many of the

Elders are gone, and others speak only broken English. Clearly the girls tend to pay strict attention to what the Elders have to say, and I am doing my best to record it as accurately as possible. Many of the boys have also been helpful.

One of the most intriguing stories the boys have told me is about the time, several years ago, that they found some bones which they could not identify. The bones were very thick, and they were puzzled so they brought one of the bones back to the village and sent it off to a university. I suspect that the bone was sent to the University of Northern British Columbia. They are not sure where the bone ended up. From their description, it may just be part of a skull bone of a Giant.

The only thing they know for sure is that the University called them up and asked if they had any monkeys up here. They assured the University that such was not the case. They have yet to hear any further word from that university.

For those of you who are sceptical, which pretty well includes the whole world, I can only say that I have spoken to the experts on these mountains, and by that I mean the Elders of the First Nations. They are the authorities on these mountains and the animals which live in them. I have recorded their stories as faithfully as possible. If course, the conclusions that I have drawn from those stories are strictly my own. If there is any mistake, it is on my part, because I believe every word they are telling me.

God Bless Them, One and All.